Basic Capacitor Questions

for

Marine Engineers

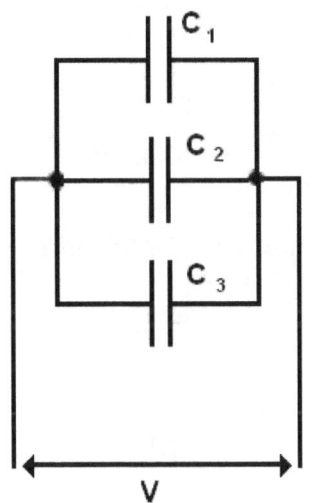

Dedication

One thing I ask from the LORD,
this only do I seek:
that I may dwell in the house of the LORD
all the days of my life,
to gaze on the beauty of the LORD
and to seek him in his temple.
For in the day of trouble
he will keep me safe in his dwelling;
he will hide me in the shelter of his sacred tent
and set me high upon a rock.

Then my head will be exalted
above the enemies who surround me;
at his sacred tent I will sacrifice with shouts of joy;
I will sing and make music to the LORD.

Hear my voice when I call, LORD;
be merciful to me and answer me.
My heart says of you, "Seek his face!"
Your face, LORD, I will seek.

(Psalm 27 vs. 4-8 Of David. NIV)

Contents

Page No.

Dedication 2

Contents 3

Foreword 5

Questions 9

Worked Solutions 33

Other Titles by the Author 104

Copyright © Christopher Lavers 2013
First Edition 2013
The Author asserts the moral right to be identified as the author of this work. All rights reserved. No part of this publication may be reproduced, stored in a retrieval system, or transmitted, in any form or by any means, electronic, mechanical, photocopying, recording or otherwise, without the prior permission of the author.

Foreword

This book is intended to develop student proficiency dealing with basic theoretical capacitor concepts underpinning Basic Electrotechnology developed in most maritime-related courses, whether Naval, Coastguard, or Merchant Marine Engineering. This book provides some 50 questions with their corresponding fully worked solutions for a range of capacitor based problems starting at a level consistent with problems covered in *Basic Electrotechnology for Marine Engineers* (by Christopher Lavers, Edmund GR Kraal, and Stanley Buyers, Volume 6 in the Reeds Marine Engineering and Technology Series ISBN: 9781408176061).

Questions and answers continue to move through an intermediate level of study, bridging the gap between *Basic Electrotechnology for*

Marine Engineers, and *Advanced Electrotechnology for Marine Engineers* (Volume 7 in the Reeds Marine Engineering and Technology Series (by Christopher Lavers, and Edmund GR Kraal). *Basic Capacitor Questions for Marine Engineers* develops Basic Electrotechnology capacitor related questions beyond the current Electrotechnology syllabi of the UK Department for Transport Examinations, and is suitable for study alongside Volume 6 as well as other volumes of the Lavers' *Basic Topics for Marine Engineers series* (lulu.com).

Knowledge regarding capacitors and other electromagnetic devices generally is now essential to merchant navy sea qualifying requirements, notably *Standards of Training, Certification and Watch keeping* (95 STCW95), as mandated by the UK Department for Transport Maritime Coastguard Agency (MCA).

This particular questions and answers volume was designed for self-study by Marine Engineers who may be at sea, and has been written in as simple a manner as possible to support the increasing number of students for whom English is not their first language of study.

Christopher Lavers

BASIC CAPACITOR QUESTIONS FOR MARINE ENGINEERS

Q1. Three capacitors C_1, C_2 and C_3 are connected in series (figure 1). By considering the potential difference across the circuit derive the total capacitance for the complete circuit.

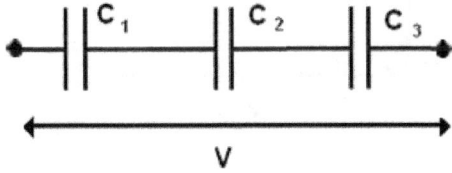

Figure 1

Q2. The same voltage is applied to each capacitor connected in parallel (figure 2). By considering the charge across the circuit derive the total capacitance for the complete circuit.

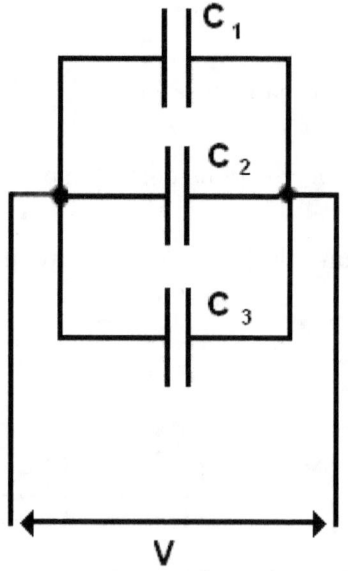

Figure 2

Q3. If two capacitors of values 200μF and 25μF respectively, are connected (a) in series (3 significant figures), and (b) in parallel (3 significant figures) across a steady applied voltage of 500 V, calculate the total circuit capacitance.

Q4. If three capacitors of values: 300 μF, 100 μF and 20 μF respectively, are connected (a) in series (3 significant figures), and (b) in parallel (3 significant figures) across a steady applied voltage of 24 V, calculate the total circuit capacitance.

Q5. How many electrons are displaced when a potential difference of 220 V exists between the parallel plates of a 6 μF capacitor (3 significant figures)?

Q6. From the relationship between Charge and Capacitance obtain an expression for the current at any instant.

Q7. Find an expression for the energy stored in an electric field and dielectric in terms of the voltage V across the capacitor.

Q8. Consider the capacitor arrangement of Question 3 and calculate the total energy stored for a steady applied voltage of 500 V, for series (3 decimal places), and parallel (3 decimal places) connections.

Q9. Consider a capacitor of 400 micro Farads capacitance. Calculate the total energy stored for a steady applied voltage of 110 V (2 decimal places).

Q10. Given that Absolute permittivity = relative permittivity $\times$ permittivity of free space

$$\text{or } \epsilon = \epsilon_r \times \epsilon_0$$

and if permeability is defined as the ratio $\dfrac{\text{Flux density}}{\text{Magnetising force}}$

$$\text{or } \mu = \frac{B}{H}$$

and permittivity = $\dfrac{\text{Electric flux density}}{\text{Electric force}}$ or $\epsilon = \dfrac{D}{E}$.

Find an expression for the permittivity in free space for a capacitor of area dimensions A and plate spacing d. For air, permittivity is measured to be $= \dfrac{1}{4\pi \times 9 \times 10^9}$ farads per metre

or $\epsilon_0 = 8.85 \times 10^{-12}$ farads per metre.

Q11. Find the capacitance of a Capacitor with Absolute permittivity ϵ and area of plates A m^2, and spacing d metres, *i.e.* the thickness of the dielectric.

Q12. A capacitor consists of 2 parallel metal plates, each 200 mm $\times$ 200 mm, separated by a sheet of polythene 3.5 mm thick, with relative permittivity 2.4. Calculate (a) the capacitance (4 decimal places) and, the energy stored in the capacitor when connected to a D.C. supply of 225 V (4 significant figures).

Q13. A capacitor consists of 2 parallel metal plates, each 250 mm × 250 mm, separated by a sheet of material 4.1 mm thick, with relative permittivity 3.1. Calculate the energy stored in the capacitor when connected to a D.C. supply of 150 V (Both 4 decimal places).

Q14. A capacitor of 4μF charged to a P.D. of 110 V is connected in *parallel* with an identical uncharged capacitor. What quantity of electricity flows into the second capacitor and to what voltage will it be charged (Both 2 significant figures)?

Q15. For a capacitor of capacitance C in a series A.C. CIRCUIT with a resistor of resistance R find an equation for the instantaneous current across the capacitor.

Q16. The voltage against time for a charging capacitor increases in a non-linear manner, following an exponential law. The equation for this line is:

$v = V(1 - e^{-t/\tau})$ volts

Where v = Instantaneous P.D. across the capacitor, V the applied circuit voltage, t the time from switch on, and $\tau = CR$ the circuit time constant. Derive an expression for the instantaneous current.

Q17. Consider a fully charged capacitor C of capacitance 2×10^{-5} Farads discharged in series through a resistor R of value 10k Ohms. The capacitor acts as a supply source as it discharges the P.D. across it, falling exponentially in value. What will be the instantaneous current at t = 2 seconds if V = 500 V (3 significant figures)?

Q18. A 25 kΩ resistor and a 15 µF capacitor are connected in series to a 220 V D.C. supply. Find the circuit current (3 decimal places) and the P.D. across the capacitor after 0.5 seconds (3 significant figures) from switch on.

Q19. A 10 kΩ resistor and a 225 µF capacitor are connected in series to a 240 V D.C. supply. Find the circuit current (3 significant figures) and the P.D. across the capacitor after 0.15 seconds (2 decimal places) from switch on.

Q20. A 30 µF capacitor, fully charged to a voltage of 200 V, is discharged through a 1 kΩ resistor. Find the time taken for the capacitor voltage to fall to 50 volts (4 decimal places).

Q21. A 22 µF capacitor, fully charged to a voltage of 210 V, is discharged through a 10 kΩ resistor. Find the time taken for the capacitor voltage to fall to 105 volts (3 decimal places).

Q22. Two capacitors of 0.01µF and 0.03 µF are connected in series across a 110 V D.C. supply. Find the voltage drop across each capacitor (1 decimal place).

Q23. Two capacitors of 0.01 µF and 0.03 µF are connected in parallel across a 110 V D.C. supply. Find the charge across each capacitor (3 significant figures).

Q24. Two capacitors of 0.02 µF and 0.06 µF are connected in series across a 3000V D.C. supply. Find the voltage drop across each capacitor (3 significant figures).

Q25. Two capacitors of 0.03 µF and 0.07 µF are connected in series across a 3000 V D.C. supply. Find the voltage drop across each capacitor (3 significant figures).

Q26. For the circuit shown, calculate the effective capacitance between A and B. The capacitance values shown are in microfarads (1 significant figure).

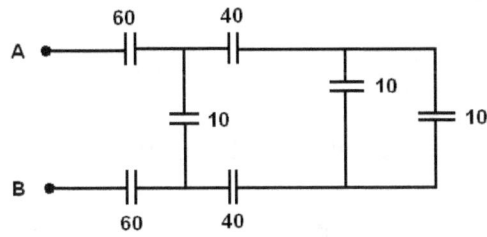

Figure 3

Q27. For the circuit shown, calculate the total equivalent effective capacitance between A and B. The capacitance values shown are in microfarads (3 decimal places).

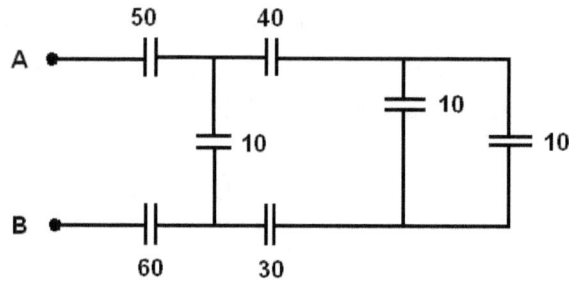

Figure 4

Q28. A variable capacitor having a capacitance of 600 µF is charged to a P.D. of 200 V. The plates of the capacitor are then separated by means of an insulating layer, so that the capacitance is reduced to 200 µF. Find, by calculation, by how much the voltage changes (1 decimal place).

Q29. A variable capacitor having a capacitance of 700 µF is charged to a P.D. of 110 V. The plates of the capacitor are then separated by means of an insulating layer, so that the capacitance is reduced to 100 µF. Find, by calculation, by how much the voltage changes (1 decimal place).

Q30. A parallel plate capacitor consists of a total of 12 metal-foil plates each 3120 mm^2 and separated by insulating spacers 0.2 mm thick. Find the capacitance of the assembly if the relative permittivity is 5 (3 significant figures).

Q31. A parallel plate capacitor consists of a total of 24 metal-foil plates each 4000 mm^2 and separated by insulating spacers 0.3 mm thick. Find the capacitance of the assembly if the relative permittivity is 7 (2 significant figures).

Q32. A P.D. of 6 kV is applied to the terminals of a capacitor consisting of two circular plates, each having an area of 5000 mm^2, separated by a dielectric 1.1 mm thick. If the capacitance is 3.5 × 10^{-4} μF, calculate the electric flux density (2 significant figure), and the permittivity of the dielectric (2 decimal places).

Q33. A P.D. of 8 kV is applied to the terminals of a capacitor consisting of two circular plates, each having an area of 9000 mm^2, separated by a dielectric 2mm thick. If the capacitance is 6 × 10^{-4} μF, calculate the electric flux density (5 significant figures), and the permittivity of the dielectric (2 decimal places).

Q34. A capacitor consists of two parallel metal plates, each 400 mm by 400 mm, separated by a sheet of polythene 4.5 mm thick, having a relative permittivity of 3.1. Calculate the energy stored in the capacitor when connected to a D.C. supply of 220 V (4 significant figures).

Q35. A capacitor consists of two parallel metal plates, each 300 mm by 300 mm, separated by a sheet of polythene 2.5 mm thick, having a relative permittivity of 3.0. Calculate the energy stored in the capacitor when connected to a D.C. supply of 110V (4 significant figures).

Q36. Calculate the capacitance value of a capacitor which has 12 parallel plates separated by insulating material 0.4 mm thick. The area of one side of each plate is 2200 mm^2 and the relative permittivity of the dielectric is 3.5 (2 significant figures).

Q37. Calculate the capacitance value of a capacitor which has 22 parallel plates separated by insulating material 0.25 mm thick. The area of one side of each plate is 1600 mm^2 and the relative permittivity of the dielectric is 6 (2 significant figures).

Q38. Two capacitors A and B having capacitances of 10 µF and 20 µF respectively are connected in series to a 800 V D.C. supply. Determine the P.D. across each capacitor (1 decimal place). If a third capacitor C is connected in *parallel* with A and it is then found that the P.D. across B is 200 V, calculate the value of C (2 significant figures), and the energy stored in it (2 decimal place).

Q39. Two capacitors A and B having capacitances of 125 µF and 45 µF respectively are connected in series to a 300 V D.C. supply. Determine the P.D. across each capacitor (3 significant figures). If a third capacitor C is connected in *parallel* with A and it is then found that the P.D. across B is 100 V, calculate the value of C (2 significant figures), and the energy stored in it (1 decimal place).

Q40. Two capacitors A and B having capacitances of 12 µF and 32 µF respectively are connected in series to a 400 V D.C. supply. Determine the P.D. across each capacitor (1 decimal place). If a third capacitor C is connected in *series* with A and B and it is then found that the P.D. across B is 300 V, calculate the value of C (2 significant figures), and the energy stored in it (1 decimal place).

Q41. Two capacitors A and B having capacitances of 10 μF and 20 μF respectively are connected in series to a 800 V D.C. supply. Determine the P.D. across each capacitor (1 decimal place). If a third capacitor C is connected in *series* with A and B and it is then found that the P.D. across B is 200 V, calculate the value of C (2 significant figures), and the energy stored in it (1 decimal place).

Q42. A D.C. voltage of 300 V is applied to a 20 μF capacitor. Find the value of the charging current at the instants when the voltage varies as follows:

Time $\left(\dfrac{1}{1000}\text{sec.}\right)$	0-1	1-2	2-3	3-4	4-5
Voltage values	0-50	50-75	80 constant	75-50	50-0

Q43. A single-phase concentric cable takes a current of 11 A per kilometre when connected to 11kV, 50 Hz mains. Calculate the capacitance of the concentric cable (2 significant figures).

Q44. A single-phase concentric cable takes a current of 12 A per kilometre when connected to 9kV, 50 Hz mains. Calculate the capacitance of the concentric cable (2 significant figures).

Q45. A coil of 200 Ω resistance and 0.2 H inductance is connected in series with a 0.35 μF capacitor to a 220 V variable frequency A.C. supply. Calculate the resonant frequency and the P.D. across the capacitor at resonance (2 decimal places).

Q46. A coil of 200 Ω resistance and 0.7 H inductance is connected in series with a 0.25 μF capacitor to a 230 V variable frequency A.C. supply. Calculate the resonant frequency and the P.D. across the capacitor at resonance (2 decimal places).

Q47. Each phase of a star-connected load consists of a resistor of 20 Ω in parallel with a 400 μF capacitor. Calculate the line current, power and power factor when the above load is connected to a 230 V, 60 Hz, three-phase supply (all 2 decimal places). What power in kW would be dissipated in the load, if it is reconnected in delta (1 decimal place).

Q48. A non-inductive coil of 5 Ω resistance is connected in parallel with an inductive coil of 7 Ω resistance and 25 Ω impedance at 60 Hz. If a potential difference of 220 V is applied to the terminals, find the current in each coil and in the mains. If a capacitor of 200 μF is connected in parallel with these coils, calculate the total current (1 decimal place).

Q49. A replacement relay coil for an alarm circuit is rated to operate from a 120 V, 50 Hz supply. It is rated 1050 Ω, 1.5 H. The coil is required to replace a damaged unit from a 220 V, 60 Hz circuit and, in order to put the coil into operation, it is decided to use a capacitor as a voltage-dropping device. Estimate the size of the ideal capacitors which should be used (2 decimal places).

Q50. Each phase of a star-connected load consists of a resistor of 10 Ω in parallel with a 400 μF capacitor. Calculate the line current, power and power factor when the above load is connected to a 230 V, 60 Hz, three-phase supply (2 decimal places). What power in kW would be dissipated in the load, if it is reconnected in delta (1 decimal place)?

BASIC CAPACITOR WORKED SOLUTIONS FOR MARINE ENGINEERS

ANSWERS

A1. *Three capacitors C_1, C_2 and C_3 are connected in series (figure 1). By considering the potential difference across the circuit derive the total capacitance for the complete circuit.*

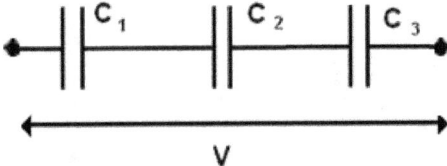

Figure 1

Let capacitors have values: C_1, C_2 and C_3 farads respectively, and the applied voltage V dropped as shown.

Then as $V = V_1 + V_2 + V_3$ and,

Since $V_1 = \dfrac{Q_1}{C_1}$ $V_2 = \dfrac{Q_2}{C_2}$ and $V_3 = \dfrac{Q_3}{C_3}$

we can write:

$$V = \dfrac{Q_1}{C_1} + \dfrac{Q_2}{C_2} + \dfrac{Q_3}{C_3}$$

If C is taken to be the equivalent capacitance of the arrangement then:

$$V = \dfrac{Q}{C}$$

or $\dfrac{Q}{C} = \dfrac{Q_1}{C_1} + \dfrac{Q_2}{C_2} + \dfrac{Q_3}{C_3} \quad \cdots (1)$

but the same current flows through each capacitor for the same time, so $Q = Q_1 = Q_2 = Q_3$ and equation (1) may be simplified to:

$$\dfrac{1}{C} = \dfrac{1}{C_1} + \dfrac{1}{C_2} + \dfrac{1}{C_3} \text{ etc.}$$

A2. *The same voltage is applied to each capacitor connected in parallel (figure 2). By considering the charge across the circuit derive the total capacitance for the complete circuit.*

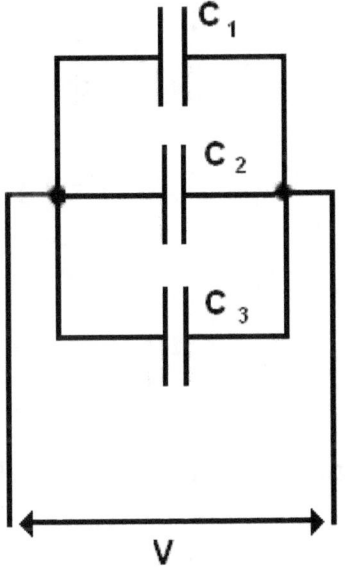

Figure 2

Then for each capacitor $Q_1 = C_1 V$, $Q_2 = C_2 V$, and $Q_3 = C_3 V$

If the total quantity of charge $= Q$ then,

$$Q = C_1 V + C_2 V + C_3 V = V(C_1 + C_2 + C_3)$$

or $\dfrac{Q}{V} = C_1 + C_2 + C_3$

If C is the equivalent capacitance of the arrangement, then;

$$Q = CV \text{ or } CV = V(C_1 + C_2 + C_3)$$

Thus $C = C_1 + C_2 + C_3$.

A3. *If two capacitors of values 200 µF and 25 µF respectively, are connected (a) in series (3 significant figures), and (b) in parallel (3 significant figures) across a steady applied voltage of 500 V, calculate the total circuit capacitance.*

(a) Series. Total circuit capacitance C is given by:

$$\frac{1}{C} = \frac{1}{200} + \frac{1}{25} = \frac{9}{200}$$

$$C = \frac{200}{9} = 22.2 \mu F$$

(b) Parallel. Total circuit capacitance is given by C = 200 + 25 or C = 225 µF.

A4. *If three capacitors of values: 300 µF, 100 µF and 20 µF respectively, are connected (a) in series (3 significant figures), and (b) in parallel (3 significant figures) across a steady applied voltage of 24 V, calculate the total circuit capacitance.*

(a) Series. Total circuit capacitance C is given by:

$$\frac{1}{C} = \frac{1}{300} + \frac{1}{100} + \frac{1}{20}$$

$$\frac{1}{C} = \frac{19}{300}$$

$$C = \frac{300}{19} = 15.8 \mu F$$

(b) Parallel. Total circuit capacitance is given by $C = 300 + 100 + 20$ or $C = 320$ µF.

A5. *How many electrons are displaced when a potential difference of 220 V exists between the plates of a 6 µF capacitor (3 significant figures)?*

$$\text{Since } Q = CV$$

then $Q = 6 \times 10^{-6} \times 220 = 1.32 \times 10^{-3}$ coulombs, but 1

$$\text{coulomb} = 6.3 \times 10^{18} \text{ electrons}$$

$$\therefore \text{ No of electrons} = 8.32 \times 10^{15}.$$

A6. *From the relationship between Charge and Capacitance obtain an expression for the current at any instant.*

From the relation Q = CV, the following can be deduced.

Since Q = It thus It = CV and,

$$I = C\frac{V}{t}$$

Current only flows whilst the voltage across a capacitor changes, as $\dfrac{V}{t}$ represents the rate of voltage change. Current at any instant can be found from the rate of change of the voltage at that instant and the capacitance.

A7. *Find an expression for the energy stored in an electric field and dielectric in terms of the voltage V across the capacitor.*

Consider the voltage to rise uniformly across the capacitor plates to a value of V volts, in a time t seconds. The average potential difference will be $\dfrac{V}{2}$ volts and the charging current constant is equal to I amperes. The average power supplied during the charging period will be $\dfrac{V}{2} \times I$ watts, and the energy stored will be:

$\dfrac{V}{2} \times I \times t$ joules. Since a capacitor has no resistance, energy is not converted into heat, but instead does work in setting up the electric field. It is this energy which is stored, and then recovered again when the electric field collapses as the capacitor is discharged.

Thus:

Energy stored = $\dfrac{V}{2} It$ joules or = $\dfrac{V}{2} Q$ joules   and

alternatively, $W = \tfrac{1}{2} CV^2$ joules.

A8. *Consider the capacitor arrangement of Question 3. Calculate the total energy stored for a steady applied voltage of 500 V, for series (3 decimal places), and parallel (3 decimal places) circuits.*

The total capacitance for each arrangement is used for C in the energy expression.

(a) Series.

Energy stored is given by $W = \frac{1}{2}CV^2$ joules

$= \frac{1}{2} \times 22.2 \times 10^{-6} \times 500^2$

$= 2.775$ J

(b) Parallel.

Energy stored or $W = \frac{1}{2} \times 225 \times 10^{-6} \times 500^2$

$= 28.125$ J.

A9. *Consider a capacitor of 400 micro Farads capacitance. Calculate the total energy stored for a steady applied voltage of 110V (2 decimal places).*

Energy stored is given by $W = \frac{1}{2}CV^2$ joules

$= \frac{1}{2} \times 400 \times 10^{-6} \times 110^2$

$= 2.42$ J

A10. *Given that Absolute permittivity = relative permittivity $\times$ permittivity of free space*

$$\text{or } \epsilon = \epsilon_r \times \epsilon_0$$

and if permeability is defined as the ratio $\dfrac{\text{Flux density}}{\text{Magnetising force}}$

$$\text{or } \mu = \frac{B}{H}$$

and permittivity = $\dfrac{\text{Electric flux density}}{\text{Electric force}}$ or $\epsilon = \dfrac{D}{E}$.

Find an expression for the permittivity in free space for a capacitor of area dimensions A and plate spacing d. For air, permittivity is measured to be = $\dfrac{1}{4\pi \times 9 \times 10^9}$ farads per metre

or $\epsilon_0 = 8.85 \times 10^{-12}$ farads per metre.

If unity is taken for the area dimensions A and d is the plate spacing:

$$\text{Then since } \epsilon_0 = \dfrac{D}{E} = \dfrac{Q/A}{V/d} = \dfrac{Q}{V} \times \dfrac{d}{A}$$

$$\text{or } \epsilon_0 = \dfrac{CV}{V} \times \dfrac{d}{A} = \dfrac{Cd}{A}$$

For a vacuum, the capacitance value of the standard capacitor, using unity for *A* and *d*, is measured to be 8.85 × 10^{-12} SI units.

or $\epsilon_0 = \dfrac{1}{4\pi \times 9 \times 10^9}$ also expressed as: 8.85×10^{-12} farads per metre.

Note. Although a vacuum is mentioned for the above capacitor arrangement, air can be taken as the dielectric, as the variation is small enough to be neglected.

A11. *Find the capacitance of a Capacitor with Absolute permittivity ϵ and area of plates A m^2, and spacing d metres, i.e. the thickness of the dielectric.*

Applied voltage is taken as V volts, resulting in a charge of Q coulombs. Charge Q is assumed to be uniformly distributed over the whole area of the plates, so the electric flux density D will be $\dfrac{Q}{A}$.

The electric force or potential gradient E in the dielectric is $\dfrac{V}{d}$ volts per metre and permittivity ϵ (by definition) $= \dfrac{D}{E}$.

$$\text{Thus } \epsilon = \dfrac{D}{E} \text{ or } \epsilon = \dfrac{Q/A}{V/d} = \dfrac{Qd}{VA}.$$

$$\text{When } \epsilon = \dfrac{CVd}{VA} = \dfrac{Cd}{A} \text{ or } C = \dfrac{\epsilon A}{d} \text{ but } \epsilon = \epsilon_0 \epsilon_r$$

$$\text{So } C = \dfrac{\epsilon_0 \epsilon_r A}{d} \text{ farads.}$$

A12. A capacitor consists of 2 parallel metal plates, each 200mm × 200mm, separated by a sheet of polythene 3.5mm thick, with relative permittivity 2.4. Calculate (a) the capacitance (4 decimal places) and, the energy stored in the capacitor when connected to a D.C. supply of 225 V (4 significant figures).

$$C = \frac{\epsilon_0 \epsilon_r A}{d}$$

So Capacitance $= \dfrac{8.85 \times 10^{-12} \times 2.4 \times (200 \times 10^{-3})^2}{3.5 \times 10^{-3}}$ farads

$C = 242.7 \times 10^{-12}$ F, or $C = 242.7$ pF

Energy stored $= \frac{1}{2} CV^2$ joules

$= \frac{1}{2} \times 242.7 \times 10^{-12} \times 225^2$

$= 6.143 \times 10^{-6}$ joules

$= 6.143$ µJ

A13. *A capacitor consists of 2 parallel metal plates, each 250 mm × 250 mm, separated by a sheet of material 4.1 mm thick, with relative permittivity 3.1. Calculate the energy stored in the capacitor when connected to a D.C. supply of 150 V (Both 4 decimal places).*

$$C = \frac{\epsilon_0 \epsilon_r A}{d}$$

So Capacitance $= \dfrac{8.85 \times 10^{-12} \times 3.1 \times (250 \times 10^{-3})^2}{4.1 \times 10^{-3}}$ farads

$C = 4.182 \times 10^{-10}$ F, or $C = 418.2$ pF

Energy stored $= \frac{1}{2} CV^2$ joules

$= \frac{1}{2} \times 4.182 \times 10^{-10} \times 150^2$

$= 4.705 \times 10^{-6}$ joules

$= 4.705$ μJ

A14. *A capacitor of 4 µF charged to a P.D. of 110 V is connected in parallel with an identical uncharged capacitor. What quantity of electricity flows into the second capacitor and to what voltage will it be charged (both 2 significant figures)?*

Consider the first capacitor designated A, then as $Q = C_A V$,

$Q = 4 \times 10^{-6} \times 110 = 4.4 \times 10^{-4}$ coulombs.

When capacitor B is connected across A, charge passes from A to B until the potential of each is the same. The capacitor arrangement is considered as a **parallel** connection or the combined circuit capacitance is the same as that of 1 capacitor with a capacitance of 8 µF since:

$C = C_1 + C_2 = 4 + 4 = 8$ µF

Applying the formula $Q = CV$

Then: $V = \dfrac{Q}{C} = \dfrac{4.4 \times 10^{-4}}{8 \times 10^{-6}}$

or $V = 55$ volts, the final voltage.

A15. For a capacitor of capacitance *C* in a series A.C. CIRCUIT with a resistor of resistance *R* find an equation for the instantaneous current across the capacitor.

$V = $ P.D. across R + P.D. across *C*

$V = $ iR + *v* where *v* is the instantaneous voltage.

$$\therefore i = \dfrac{V - v}{R}$$

At the instant of switching the current on, the instantaneous voltage value *v* across the capacitor will be zero.

Consequently the current (i) will be at its *maximum*, limited only by the resistance R.

At switch on $i = \dfrac{V}{R}$, but this current is the charging capacitor current and, as the capacitor charges, v increases, which in turn decreases i. The rate of charging therefore decreases until, when the capacitor is fully charged, current is zero. However, zero current is achieved after an infinite charging time but, for practical purposes, it is assumed this occurs in a time equal to 5 times the initial charging rate. If the initial charging rate is maintained, a capacitor will now be fully charged in a fixed time depending upon the specific resistor and capacitance circuit components. This charging time is the *Time Constant* and given the symbol τ (Greek letter TAU) where: $\tau = CR$ seconds.

A16. *The voltage against time for a charging capacitor increases in a non-linear manner, following an exponential law. The equation for this line is:*

$v = V(1 - e^{-t/\tau})$ *volts where v = Instantaneous P.D. across the capacitor, V the applied circuit voltage, t the time from switch on, and $\tau = CR =$ the circuit time constant.*

Derive an expression for the instantaneous current.

The equation for this curve is:

$i = I e^{-t/\tau}$ or more generally $i = \dfrac{V}{R} e^{-t/\tau}$ when t = 0

and $i = I$

At t = 0 the instantaneous voltage will be zero, and current I will be at its maximum, limited only by the resistance so I = V/R .

A17. *Consider a fully charged capacitor C of capacitance 2×10^{-5} Farads discharged in series through a resistor R of value 10 k Ohms. The capacitor acts as a supply source as it discharges the P.D. across it, falling exponentially in value. What will be the instantaneous current at t = 2 seconds if V = 500 V (3 significant figures?)*

As the current is limited by resistance (I = *v*/R at the start of the discharge the current is maximum) so

I = V/R

Hence on discharge $v = Ve^{-t/\tau}$ and $i = e^{-t/\tau}$

$i = V/R \, e^{-t/\tau}$ $RC = 0.2$

$i = 500/10000 \, e^{-2/(RC)} = 5/100 \, e^{-2/(0.2)}$

$i = 2.27 \times 10^{-6}$ A.

A18. *A 25 kΩ resistor and a 15 μF capacitor are connected in series to a 220 V D.C. supply. Find the circuit current (3 decimal places) and the P.D. across the capacitor after 0.5 seconds (3 significant figures) from switch on.*

Time Constant $\tau = CR = 15 \times 10^{-6} \times 25 \times 10^{3}$

$= 0.375$ s.

$$i = \frac{V}{R} e^{-t/\tau} = \frac{220}{25\times 10^3} e^{-0.5/0.375}$$

$I = 8.8 \times 10^{-3} \times e^{-0.5/0.375}$

$$= 2.32 \times 10^{-3} \text{ A}$$

$$v = 220(1 - e^{-0.5/0.375})$$

$$= 162 \text{ V}$$

A19. *A 10 kΩ resistor and a 225 µF capacitor are connected in series to a 240 V D.C. supply. Find the circuit current (3 significant figures) and the P.D. across the capacitor after 0.15 seconds (2 decimal places) from switch on.*

$$\text{Time Constant } \tau = CR$$

$$= 225 \times 10^{-6} \times 10 \times 10^{3}$$

$$= 2.25 \text{ s.}$$

$$i = \frac{V}{R} e^{-t/\tau} = \frac{240}{10 \times 10^{3}} e^{-0.15/2.25}$$

$$= 0.0225 \text{ A}$$

$$v = V(1-e^{-0.15/2.25})$$

$$= 240(1-e^{-0.15/2.25})$$

$$= 15.48 \text{ A}$$

A20. *A 30 µF capacitor, fully charged to a voltage of 200 V, is discharged through a 1 kΩ resistor. Find the time taken for the capacitor voltage to fall to 50 volts (4 decimal places).*

$$\tau = CR = 30 \times 10^{-6} \times 1 \times 10^3$$

$$= 0.03 \text{ s}$$

$$v = Ve^{-t/\tau} = 200e^{-t/0.03}$$

$$50 = 200e^{-t/0.03}$$

$$50/200 = e^{-t/0.03}$$

$$= 0.25$$

Taking logs to base e.

$$-t/0.03 = -0.60206 \text{ so } t = 0.0181 \text{ s}$$

A21. *A 22 µF capacitor, fully charged to a voltage of 210 V, is discharged through a 10 kΩ resistor. Find the time taken for the capacitor voltage to fall to 105 volts (3 decimal places).*

$$\tau = CR$$

$$= 22 \times 10^{-6} \times 10 \times 10^{3}$$

$$= 0.22 \text{ s}$$

$$v = V e^{-t/\tau}$$

$$105 = 210\, e^{-t/0.22}$$

$$105/210 = e^{-t/0.22}$$

$$= 0.5$$

Taking logs to base e.

$$-t/0.22 = -0.30103$$

$$t = 0.066 \text{ s}$$

A22. *Two capacitors of 0.01 µF and 0.03 µF are connected in series across a 110 V D.C. supply. Find the voltage drop across each capacitor (1 decimal place).*

For a series combination, the equivalent capacitance is given by C, where: $\dfrac{1}{C} = \dfrac{1}{0.01} + \dfrac{1}{0.03}$ (in microfarads)

So $C = 7.5$ micro Farads Also $Q = CV$

∴ $Q = 7.5 \times 10^{-3} \times 110 = 0.825$ micro coulombs

Then $V_1 = \dfrac{0.825 \times 10^{-6}}{0.01 \times 10^{-6}} = 82.5$ V

and $V_2 = \dfrac{0.825 \times 10^{-6}}{0.03 \times 10^{-6}} = 27.5$ V

The voltage drops are respectively 82.5 V and 27.5 V

A23. *Two capacitors of 0.01μF and 0.03μF are connected in parallel across a 110 V D.C. supply. Find the charge across each capacitor (3 significant figures).*

For a parallel combination, the equivalent capacitance is given by C where: $C = C_1 + C_2$

$= 0.01 + 0.03 = 0.04$ in microfarads.

C = Q/V Therefore Q = CV, and V = 110V

So $Q_1 = \dfrac{C_1}{V} = \dfrac{0.01 \times 10^{-6}}{110} = 9.09 \times 10^{-11}$ C

And $Q_2 = \dfrac{C_2}{V} = \dfrac{0.01 \times 10^{-6}}{110} = 2.72 \times 10^{-10}$ C

A24. *Two capacitors of 0.02 µF and 0.06 µF are connected in series across a 3000 V D.C. supply. Find the voltage drop across each capacitor (3 significant figures).*

For a series combination, the equivalent capacitance is given by C, where: $\dfrac{1}{C} = \dfrac{1}{0.02} + \dfrac{1}{0.06}$ (in microfarads)

So $C = 0.015$ µF Also $Q = CV$

∴ $Q = 0.015 \times 10^{-6} \times 3000 = 4.5 \times 10^{-3}$ coulombs

Then $V_1 = \dfrac{4.5 \times 10^{-3}}{0.02 \times 10^{-6}} = 2250$ V and $V_2 = \dfrac{4.5 \times 10^{-3}}{0.06 \times 10^{-6}} = 750$ V

The voltage drops are respectively 2250 V and 750 V

A25. Two capacitors of 0.03 µF and 0.07 µF are connected in series across a 3000V D.C. supply. Find the voltage drop across each capacitor (3 significant figures).

For a series combination, the equivalent capacitance is given by:

$$\dfrac{1}{C} = \dfrac{1}{0.02} + \dfrac{1}{0.06}$$

= 0.021 micro farads or 0.021 µF

C = Q/V Therefore Q = CV, and V = 110 V

∴ $Q = 0.021 \times 10^{-6} \times 3000 = 6.3 \times 10^{-5}$ Coulombs

Then $V_1 = \dfrac{6.3 \times 10^{-5}}{0.03 \times 10^{-6}} = 2100$ V

And $V_2 = \dfrac{6.3 \times 10^{-5}}{0.07 \times 10^{-6}} = 900$ V

The voltage drops are respectively 2100 V and 900 V

A26. *For the circuit shown, calculate the effective capacitance between A and B. The capacitance values shown are in microfarads (1 significant figure).*

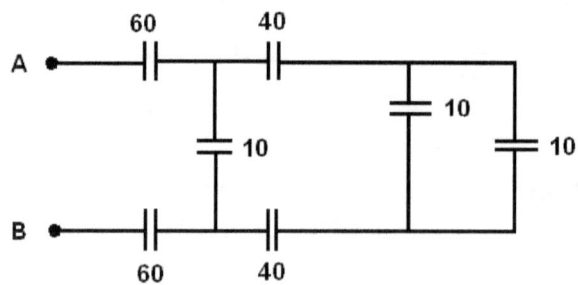

Figure 3

The final two parallel 10 μF capacitors are equivalent to a single capacitor of value 20 μF, (C = 10 μF + 10 μF).

The capacitance C of the middle section consists of: 40 μF, 20 μF and 40 μF in series given by:

$$\frac{1}{C} = \frac{1}{40} + \frac{1}{20} + \frac{1}{40} \quad \text{or } C = 10 \ \mu F$$

This series circuit is in parallel with a 10 μF capacitor with an equivalent capacitance

C = 10 μF + 10 μF = 20 μF.

The final arrangement between A and B is equivalent to a 60 μF, 20 μF and 60 μF capacitor in series. The total equivalent capacitance is given by:

$$\frac{1}{C} = \frac{1}{60} + \frac{1}{20} + \frac{1}{60} = \frac{5}{60}$$

so C = 60/5 = 12 μF

A27. *For the circuit shown, calculate the total effective capacitance between A and B. The capacitance values shown are in microfarads (3 decimal places).*

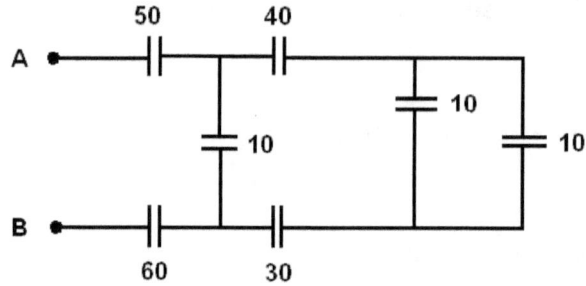

Figure 4

The final two parallel 10 μF capacitors are equivalent to a single capacitor of value 20 μF, (C = 10 μF + 10 μF).

The capacitance C of the middle section consists of: 40 μF, 20 μF and 30 μF in series given by:

$$\frac{1}{C} = \frac{1}{40} + \frac{1}{20} + \frac{1}{30} = \frac{13}{120} \text{ or } C = 120/13 = 9.2308 \ \mu F$$

This series circuit is in parallel with a 10 μ F capacitor with an equivalent capacitance C = 10 +9.2308 μ F = 19.2308 μ F.

The final arrangement between A and B is equivalent to a 50 μ F, 19.2308 μ F and 60 μ F capacitor in series. The total equivalent capacitance is given by:

$$\frac{1}{C} = \frac{1}{50} + \frac{1}{19.2308} + \frac{1}{60}$$

or $C = 11.278$ μ F

A28. A *variable capacitor having a capacitance of 600 µF is charged to a P.D. of 200 V. The plates of the capacitor are then separated by means of an insulating layer, so that the capacitance is reduced to 200 µF. Find, by calculation, by how much the potential difference changes (1 decimal place).*

Since $Q = CV$. ∴ the quantity of electricity received *initially* is given by:

$Q = 600 \times 10^{-6} \times 200 = 0.12$ Coulombs.

Since the plates are separated by an insulating layer there is no loss of charge and hence Q remains the same.

Under the new condition since, as before, $Q = CV$

Then $V = \dfrac{Q}{C} = \dfrac{0.12}{400 \times 10^{-6}} = 300\,\text{V}$

Hence the potential difference will increase by 300−200 = 100 V

A29. *A variable capacitor having a capacitance of 700 μF is charged to a P.D. of 110 V. The plates of the capacitor are then separated by means of an insulating layer, so that the capacitance is reduced to 100 μF. Find, by calculation, by how much the voltage changes (1 decimal place).*

Since $Q = CV$. ∴ the quantity of electricity received *initially* is given by:

$Q = 700 \times 10^{-6} \times 200 = 0.077$ Coulombs

Since the plates are separated by an insulating layer there is no loss of charge and hence Q remains the same.

Under the new condition since, as before, $Q = CV$

Then $V = \dfrac{Q}{C} = \dfrac{0.077}{100 \times 10^{-6}} = 770\text{V}$

Hence the potential difference will increase by 770 −110 = 660 V

A30. *A parallel plate capacitor consists of a total of 12 metal-foil plates each 3120 mm^2 and separated by insulating spacers 0.2 mm thick. Find the capacitance of the assembly if the relative permittivity is 5 (3 significant figures).*

The capacitor is made from 12 plates in parallel, 6 in one assembly, interleaved with 6 opposite plates in parallel forming the other plate assembly. There will be 11 electric fields and the total capacitance will be 11 times the capacitance between one pair of plates.

Thus C of one pair of plates $= \dfrac{\epsilon A}{d} = \dfrac{\epsilon_o \epsilon_r A}{d}$

or $C = \dfrac{8.85 \times 10^{-12} \times 5 \times 3120 \times 10^{-6}}{0.2 \times 10^{-6}}$

$= 6.903 \times 10^{-10}$ Farads

or with 11 units in parallel $C = 11 \times 6.903 \times 10^{-10}$ farads

$= 7.59 \times 10^{-9}$ F

A31. A plate capacitor consists of a total of 24 metal-foil plates each 4000 mm^2 and separated by insulating spacers 0.3 mm thick. Find the capacitance of the assembly if the relative permittivity of the insulating spacers is 7 (2 significant figures).

The capacitor is made from 24 plates in parallel, 12 in one assembly, interleaved with 12 plates in parallel forming the other plate assembly. There will be 23 electric fields and the total capacitance will be 23 times the capacitance between one pair of plates.

Thus C of one pair of plates $= \dfrac{\epsilon A}{d} = \dfrac{\epsilon_o \epsilon_r A}{d}$

or $C = \dfrac{8.85 \times 10^{-12} \times 7 \times 4000 \times 10^{-6}}{0.3 \times 10^{-6}} = 8.26 \times 10^{-6}$ F

or with 23 units in parallel C = 23 × 8.26 × 10^{-10} F = 1.9 × 10^{-8} F

A32. *A P.D. of 6 kV is applied to the terminals of a capacitor consisting of two circular plates, each having an area of 5000 mm^2, separated by a dielectric 1.1mm thick. If the capacitance is 3.5×10^{-4} µF, calculate the electric flux density (2 significant figures), and the permittivity of the dielectric (2 decimal places).*

Since $Q = CV$, then $Q = 3.5 \times 10^{-4} \times 10^{-6} \times 6 \times 10^{3}$

$= 2.1 \times 10^{-6}$ coulombs

$\therefore$ Flux density, $D = \dfrac{Q}{A} = \dfrac{2.1 \times 10^{-6}}{5000 \times 10^{-6}}$

$= 4.2 \times 10^{-4}$ coulomb per m^2

Also, permittivity, $\epsilon = \dfrac{\text{electricity flux density}}{\text{electric force}} = \dfrac{D}{E}$

And electric force, $= \dfrac{V}{d} = \dfrac{6 \times 10^3}{1.1 \times 10^{-3}} = 5454545.5 =$ 5454545.5 volts m^{-1}

So $\varepsilon_r = \dfrac{\varepsilon}{\varepsilon_0} = \dfrac{4.2 \times 10^{-4}}{5454545.5 \times 8.85 \times 10^{-12}}$ as $\epsilon = \epsilon_0 \times \epsilon_r$

and $\epsilon_r = 8.70$

A33. A P.D. of 8 kV is applied to the terminals of a capacitor consisting of two circular plates, each having an area of 9000 mm^2, separated by a dielectric 2 mm thick. If the capacitance is 6 × 10^{-4} μF, calculate the electric flux density (5 significant figures), and the permittivity of the dielectric (2 decimal places).

Since $Q = CV$, then $Q = 6 \times 10^{-4} \times 10^{-6} \times 8 \times 10^3$

= 4.8 × 10^{-6} coulombs

∴ Flux density, $D = \dfrac{Q}{A} = \dfrac{4.8 \times 10^{-6}}{9000 \times 10^{-6}}$

= 5.3333 × 10^{-4} coulomb per m^2

Also, permittivity, $\epsilon = \dfrac{\text{electricity flux density}}{\text{electric force}} = \dfrac{D}{E}$

And electric force $E = \dfrac{V}{d} = \dfrac{8 \times 10^{-3}}{2 \times 10^{-3}}$ = 4 × 10^6 volts m^{-1}

Hence $\varepsilon = \dfrac{5.3333 \times 10^{-4}}{4 \times 10^6}$ also $\epsilon = \epsilon_o \times \epsilon_r$

And $\varepsilon_r = \dfrac{\varepsilon}{\varepsilon_0} = \dfrac{5.3333 \times 10^{-4}}{4 \times 10^6 \times 8.85 \times 10^{-12}}$

or ϵ_r = 15.07

A34. A capacitor consists of two parallel metal plates, each 400 mm by 400 mm, separated by a sheet of polythene 4.5 mm thick, having a relative permittivity of 3.1. Calculate the energy stored in the capacitor when connected to a D.C. supply of 220 V (4 significant figures).

$C = \dfrac{\epsilon A}{d}$ $A = 16 \times 10^4 \times 10^{-6}$ m^2 $= 16 \times 10^{-2}$ m^2

$d = 4.5 \times 10^{-3}$ m

and $\epsilon = \epsilon_o \times \epsilon_r$ $= 8.85 \times 10^{-12} \times 3.1$

Hence $C = \dfrac{8.85 \times 10^{-12} \times 3.1 \times 16 \times 10^{-2}}{4.5 \times 10^{-3}}$ C

$= 9.7547 \times 10^{-10}$ F

Energy, $W = \frac{1}{2} CV^2$ joules

$= \frac{1}{2} \times 9.7547 \times 10^{-10} \times 220^2$

$= 2.3601 \times 10^{-5}$ joules $= 0.2361 \; \mu\mathrm{J}$

A35. A capacitor consists of two parallel metal plates, each 300 mm by 300 mm, separated by a sheet of polythene 2.5 mm thick, having a relative permittivity of 3.0. Calculate the energy stored in the capacitor when connected to a D.C. supply of 110 V (4 significant figures).

$C = \dfrac{\epsilon A}{d}$ $A = 9 \times 10^4 \times 10^{-6} \text{ m}^2 = 9 \times 10^{-2} \text{ m}^2$

$d = 2.5 \times 10^{-3}$ m

and $\epsilon = \epsilon_o \times \epsilon_r = 8.85 \times 10^{-12} \times 3.0$

$$C = \frac{8.85 \times 10^{-12} \times 3.0 \times 9 \times 10^{-2}}{2.5 \times 10^{-3}} = 6.8271 \times 10^{-3} \, C$$

$= 6.8271 \times 10^{-3}$ F

Energy, $W = \frac{1}{2} CV^2$ joules

$= \dfrac{1}{2} \times 6.8271 \times 10^{-3} \times 110^2$

$= 4.1304 \times 10^{-6}$ joules $= 4.130 \, \mu$ J

A36. Calculate the capacitance value of a capacitor which has 12 parallel plates separated by insulating material 0.4 mm thick. The area of one side of each plate is 2200 mm^2 and the relative permittivity of the dielectric is 3.5 (2 significant figures).

A 12-plate capacitor is made from two 6-plate assemblies interleaved with each other and separated by the dielectric. There are thus 11 electric fields or the final capacitance is 11 times that of one plate arrangement.

$$C = \frac{\varepsilon A}{d} \text{ where } A = 2200 \times 10^{-6} \text{ m}^2$$

$d = 0.4 \times 10^{-3}$ m

$\epsilon = \epsilon_o \times \epsilon_r$

$$C = \frac{8.85 \times 10^{-12} \times 3.5 \times 2200 \times 10^{-6}}{0.4 \times 10^{-3}} = 1.7036 \times 10^{-10}$$

$= 1.7036 \times 10^{-4}$ microfarads

Total capacitance $= 11 \times 1.8 \times 10^{-4}$ $= 0.0019 \ \mu\text{F}$

A37. *Calculate the capacitance value of a capacitor which has 22 parallel plates separated by insulating material 0.25 mm thick. The area of one side of each plate is 1600 mm^2 and the relative permittivity of the dielectric is 6 (2 significant figures).*

A 22-plate capacitor is made from two 11-plate assemblies interleaved with each other and separated by the dielectric. There are thus 21 electric fields or the final capacitance is 21 times that of one plate arrangement.

$$C = \frac{\varepsilon A}{d} \text{ where } A = 1600 \times 10^{-6} \text{ m}^2$$

$d = 0.25 \times 10^{-3}$ m

$\epsilon = \epsilon_o \times \epsilon_r$

$$C = \frac{8.85 \times 10^{-12} \times 6 \times 1600 \times 10^{-6}}{0.25 \times 10^{-3}} = 3.3984 \times 10^{-10}$$

$= 3.3984 \times 10^{-10}$ farads

Total capacitance $= 21 \times 3.3984 \times 10^{-10} = 0.071 \, \mu F$

A38. *Two capacitors A and B having capacitances of 10 μF and 20 μF respectively are connected in series to a 800 V D.C. supply. Determine the P.D. across each capacitor (3 significant figures). If a third capacitor C is connected in parallel with A and it is then found that the P.D. across B is 200 V, calculate the value of C (3 significant figures), and the energy stored in it (2 decimal place).*

Let C = capacitance of the series arrangement,

then $\dfrac{1}{C} = \dfrac{1}{10} + \dfrac{1}{20}$ or $C = 6.666\ \mu F$

The charge stored is given by $Q = CV = 6.666 \times 10^{-6} \times 800$

$= 5.3328 \times 10^{-3}$ coulombs.

P.D. across 10 μF capacitor A = $\dfrac{V}{d} = \dfrac{5.3328 \times 10^{-3}}{10 \times 10^{-6}} = 533.3$

= 533 V

P.D. across 20 µF capacitor B = 800 − 533 = 267 V

If P.D. across B is 500 V then P.D. across the parallel arrangement will be 300 V. The equivalent capacitance must be 5/3 of Capacitor B's value since the voltage is 3/5 of the value of capacitor B. So the equivalent capacitance must be 33.3 μ F.

So 33.3 μ F = 10 μ F + C

Therefore C = 23.3 μ F, being in parallel with A.

Also the energy stored, W = $\frac{1}{2}CV^2$

= $\frac{1}{2}$ × 23.3 × 10^{-6} × 300^2 joules. Thus W = 1.05 J

A39. *Two capacitors A and B having capacitances of 125 μF and 4 5μF respectively are connected in series to a 300 V D.C. supply. Determine the P.D. across each capacitor (1 decimal place). If a third capacitor C is connected in parallel with A and it is then found that the P.D. across B is 100 V, calculate the value of C (2* significant figures), and the energy stored in it (1 decimal place).

Let C = capacitance of the series arrangement,

then $\dfrac{1}{C} = \dfrac{1}{125} + \dfrac{1}{45}$ $C = 33.1 \ \mu\text{F}$

The charge stored is given by:

$Q = CV$

$= 33.1 \times 10^{-6} \times 600$

$= 0.0198529$ coulombs.

P.D. across 125 μ F capacitor $A = \dfrac{0.0198529}{125 \times 10^{-6}} = 158.8$ V

P.D. across 45 µF capacitor B = 600 − 158.8 = 441.2 V

If P.D. across B is 500 V then P.D. across the parallel arrangement will be 100 V. The equivalent capacitance must be five times the capacitance of capacitor B since the voltage is 1/5 that across B.

So the equivalent capacitance must be 5 × 45 µF = 225 µF being in parallel with A.

So 225 µF = 125 µF + C

Therefore C = 110 µF being in parallel with A.

Also the energy stored, $W = \frac{1}{2}CV^2$

$= \frac{1}{2} \times 110 \times 10^{-6} \times 100^2$ joules. Thus $W = 0.55$ J

A40. Two capacitors A and B having capacitances of 12 μF and 32 μF respectively are connected in series to a 400 V D.C. supply. Determine the P.D. across each capacitor (3 significant figures). If a third capacitor C is connected in series with A and B and it is then found that the P.D. across B is 300 V, calculate the value of C (2 significant figures), and the energy stored in it (2 decimal place).

Let C = capacitance of the series arrangement,

then $\dfrac{1}{C} = \dfrac{1}{12} + \dfrac{1}{32}$ or $C = 8.727 \; \mu F$

The charge stored is given by:

$Q = CV$

$= 8.727 \times 10^{-6} \times 400$

$= 3.4909 \times 10^{-3}$ coulombs.

P.D. across 12 μ F capacitor A $= \dfrac{3.4909 \times 10^{-3}}{12 \times 10^{-6}} = 291$ V

P.D. across 30µF capacitor B = 400 − 291 = 109 V

If P.D. across B is 300 V then P.D. across the parallel arrangement will be 100V. The equivalent capacitance must be three times the capacitance of capacitor B since the voltage is 1/3 the voltage across B.

So the equivalent capacitance must be 3 × 30 µF = 90 µF.

So 90 µF = 12 µF + C

Therefore C = 78 μ F being in parallel with A.

Also the energy stored, W = $\tfrac{1}{2}CV^2$

$= \tfrac{1}{2} \times 78 \times 10^{-6} \times 100^2$ joules. Thus $W = 0.39$ J

A41. Two capacitors A and B having capacitances of 10 μF and 20 μF respectively are connected in series to a 800 V D.C. supply. Determine the P.D. across each capacitor (1 decimal place). If a third capacitor C is connected in series with A and B and it is then found that the P.D. across B is 200 V, calculate the value of C (2 significant figures), and the energy stored in it (2 decimal place).

Let C = capacitance of the series arrangement,

then $\dfrac{1}{C} = \dfrac{1}{10} + \dfrac{1}{20}$ or $C = 6.666 \; \mu F$

The charge stored is given by $Q = CV = 6.666 \times 10^{-6} \times 800$

$= 5.3328 \times 10^{-3}$ coulombs.

P.D. across 10 μF capacitor A $V = \dfrac{Q}{C} = \dfrac{5.3328 \times 10^{-3}}{10 \times 10^{-6}} =$ 533.3V

P.D. across 20 µF capacitor B = 600 − 533.3 = 66.7 V

If P.D. across B is 200 V then P.D. across the parallel arrangement will be 200 V. The equivalent capacitance must be 20 μF as the voltage is the same as that across capacitor B.

So 20 µF = 10 µF + C

Therefore C = 10 μF being in parallel with A.

Also the energy stored, W = $\tfrac{1}{2}CV^2$

= $\tfrac{1}{2}$ × 10 × 10^{-6} × 200^2 joules. Thus $W = 0.2$J

A 42. A D.C. voltage of 300 V is applied to a 20 μF capacitor. Find the value of the charging current at the instants when the voltage varies as follows:

Time $\left(\dfrac{1}{1000}\text{sec.}\right)$	0-1	1-2	2-3	3-4	4-5
Voltage values	0-50	50-75	80 constant	75-50	50-0

Since $Q = CV$ and $Q = It$ then $It = CV$

or $I = C\dfrac{V}{t}$ where $V =$ the voltage change then,

i. $I = 20 \times 10^{-6} \times 50 / 1 \times 10^{-3}$ amperes $= 1$ A

ii. $I = 20 \times 10^{-6} \times 25 / 1 \times 10^{-3}$ amperes $= 0.5$ A

iii. $I = 20 \times 10^{-6} \times 0 / 1 \times 10^{-3}$ amperes $= 0$ A

iv. $I = 20 \times 10^{-6} \times 25 / 1 \times 10^{-3}$ amperes $= 0.5$A

v. $I = 20 \times 10^{-6} \times 50 / 1 \times 10^{-3}$ amperes $= 1$A

A43. *A single-phase concentric cable takes a current of 11 A per kilometre when connected to 11kV, 50 Hz mains. Calculate the capacitance of the concentric cable (2 significant figures).*

$$X_C = \frac{V}{A} = \frac{11000}{11} = 1000 \Omega$$

$$C = \frac{1}{2\pi f X_C}$$

$$C = \frac{1}{2\pi \times 50 \times 1000}$$

C= 3.2 μF

A44. *A single-phase concentric cable takes a current of 12 A per kilometre when connected to 9 kV, 50 Hz mains. Calculate the capacitance of the concentric cable (2 significant figures).*

$$X_C = \frac{V}{A} = \frac{9000}{12} = 750\,\Omega$$

$$C = \frac{1}{2\pi f X_C}$$

$$C = \frac{1}{2\pi \times 50 \times 750}$$

C= 4.2 µF

A45. *A coil of 200 Ω resistance and 0.2 H inductance is connected in series with a 0.35 μF capacitor to a 220 V variable frequency A.C. supply. Calculate the resonant frequency and the P.D. across the capacitor at resonance (2 decimal places).*

At resonance $2\pi fL = \dfrac{1}{2\pi fC}$

$\therefore f = \dfrac{1}{2\pi\sqrt{LC}}$

$$f = \dfrac{1}{2\pi\sqrt{0.2 \times 0.35 \times 10^{-6}}}$$

Resonant Frequency = 601.549 Hz = 601.55Hz.

At resonance there is *no* resultant reactance i.e. R = Z

$$\therefore I = \dfrac{V}{R} = \dfrac{220}{200}$$

= 1.1 A

$$X_C = \frac{1}{2\pi fC} = \frac{1}{2 \times \pi \times 601.549 \times 0.35 \times 10^{-6}} = 755.9291 \Omega$$

= 755.93 Ω

P.D. across capacitor = $IX = 1.1 \times 755.9291 = 831.52 \, V$

Q46. A coil of 200 Ω resistance and 0.7 H inductance is connected in series with a 0.25 μF capacitor to a 230 V variable frequency A.C. supply. Calculate the resonant frequency and the P.D. across the capacitor at resonance (2 decimal places).

At resonance $2\pi fL = \dfrac{1}{2\pi fC} \quad \therefore f = \dfrac{1}{2\pi\sqrt{LC}}$

$$f = \frac{1}{2\pi\sqrt{0.7 \times 0.25 \times 10^{-6}}} =$$

Resonant Frequency = 380.45 Hz

At resonance there is *no* resultant reactance i.e. R = Z

$$I = \frac{V}{R} = \frac{230}{200}$$

I = 1.15 A

$$X_C = \frac{1}{2\pi fC} = \frac{1}{2 \times \pi \times 380.45 \times 0.25 \times 10^{-6}} = 1673.33\,\Omega$$

P.D. across capacitor = $IX = 1.15 \times 1673.33 = 1924.33\,\text{V}$

A47. *Each phase of a star-connected load consists of a resistor of 20 Ω in parallel with a 400 μF capacitor. Calculate the line current, power and power factor when the above load is connected to a 230 V, 60 Hz, three-phase supply (2 decimal places). What power would be dissipated in the load, if it is reconnected in delta (1 decimal place)?*

Reactance X_C of capacitor

$$= X_C = \frac{1}{2\pi f C} = \frac{1}{2 \times \pi \times 60 \times 400 \times 10^{-6}} = 6.63\,\Omega$$

Since the load is balanced:

The voltage across a phase, $V_{ph} = \dfrac{230}{\sqrt{3}} = 132.79\,\text{V}$

Current in the resistor $I_R = \dfrac{132.79}{20} = 6.64\,\text{A}$

= 6.64 A, in phase with V_{ph}

Current in the capacitor $I_C = \dfrac{132.79}{6.63} = 20.03\,\text{A}$

= 20.03 A, leading V_{ph} by 90°

Let I_{ph} = the resultant of 6.64 A and 20.03 A which are in quadrature

$$\therefore I_{ph} = \sqrt{6.64^2 + 20.03^2} = 21.10\,\text{A}$$

= 21.10 A. This is also the line current since the load is star connected ∴ I = 21.10 A

Power factor of load = $\cos \phi = \dfrac{I_R}{I_{ph}} = \dfrac{6.64}{21.10} = 0.31$

Power of load, $P = \sqrt{3}\ VI \cos \phi$

$= \sqrt{3} \times 230 \times 21.10 \times 0.31 = 2605.7$ watts = 2.61 kW

If the load is in delta, the current per phase = $\sqrt{3}$ × original I_{ph}

$= \sqrt{3} \times 21.10$ A

The line current would be $\sqrt{3}$ times this new phase current.

∴ New I $= \sqrt{3} \times \sqrt{3} \times 21.10 = 63.3$ A

The power factor of the load will remain the same

So new power, $P = \sqrt{3}\ VI \cos \phi$

=New power

$$= \sqrt{3} \times IV = \sqrt{3} \times 230 \times 63.3 \times 0.31 = 7.8\,\text{kW}$$

Q48. A non-inductive coil of 5 Ω resistance is connected in parallel with an inductive coil of 7 Ω resistance and 25 Ω impedance at 60 Hz. If a potential difference of 220 V is applied to the terminals, find the current in each coil and in the mains. If a capacitor of 200 μF is connected in parallel with these coils, calculate the total current (1 decimal place).

Let non-inductive coil of 25 Ω be designated branch A.

Then $I_A = \dfrac{V}{R} = \dfrac{220}{5} = 44\,\text{A}$

$$\cos\phi_A = \frac{5}{5} = 1 \quad \sin\phi = 0$$

Let inductive coil of impedance 7Ω be designated branch B

Then $I_B = \dfrac{V}{R} = \dfrac{220}{25} = 8.8$ A

$\cos\phi_B = \dfrac{7}{25} = 0.28 \quad \sin\phi = 0.96$

Resolving in active and reactive components

$Ia = I_A \cos\phi_A + I_B \cos\phi_B =$
$44 \times 1 + 8.8 \times 0.28 = 44 + 2.464 = 46.464$

and

$Ir = -I_A \sin\phi_A - I_B \sin\phi_B =$
$-44 \times 0 - 8.8 \times 0.96 = -8.45$

$\therefore I = \sqrt{46.6^2 + 8.45^2} = 47.4$

Thus current taken from mains is 47.4 A

With a capacitor of 200 µF connected in parallel:

$$X_C = \frac{1}{2\pi f C} = \frac{1}{2 \times \pi \times 60 \times 200 \times 10^{-6}} = 13.3\,\Omega$$

Current $I_C = \dfrac{110}{13.3} = 16.54\,\text{A}$

Resolving as before, active component I_a remains the same but the reactive component Ir is given by:

$Ir = 16.54 - 8.45 = 8.09$ The 8.09A is now vertically upwards, *i.e.* leading *V* by 90°.

Resultant current $I = \sqrt{46.6^2 + 8.09^2} = 47.3\,\text{A}$

A49. *A replacement relay coil for an alarm circuit is obtainable but is rated to operate from a 120 V, 50 Hz supply. It is rated 1050 Ω, 1.5 H. The coil is required to replace a damaged unit from a 220 V, 60 Hz circuit and, in order to put the coil into operation, it is decided to use a capacitor as a voltage-dropping device. Estimate the size of ideal capacitor which should be used (2 decimal places).*

Reactance of new coil on 50 Hz = $X_L = 2\pi \times f \times L$

= 3.14 × 150 = 471 Ω

Impedance of new coil on 50 Hz

$$= \sqrt{1050^2 + 471^2} = 1150.8\,\Omega$$

Current taken by coil $= \dfrac{120}{1150.8} = 0.104\,\text{A}$

Required impedance on 220 V circuit = $\dfrac{220}{0.104}$ = 2115 Ω

The reactance of the 220 V circuit would be

$$X_L = \sqrt{(2115)^2 - (1050)^2} = 1836\,\Omega$$

Reactance of new coil on 50 Hz is 471 Ω

On 60 Hz it is 471 × $\frac{6}{5}$ = 565.2 Ω

Now the reactance of the required capacitor must cancel this inductive reactance and provide the additional reactance for the 220 V circuit, *i.e.* it must be X_c = 1836 + 565.2 = 2401.2 Ω.

The required value of capacitance is given by:

$$X_C = \frac{1}{2\pi f C}$$

$$X_C = \frac{1}{2\pi \times 60 \times C}$$

$$C = \frac{1}{2\pi \times 60 \times 2401.2} = 1.1 \mu F$$

A50. *Each phase of a star-connected load consists of a resistor of 10 Ω in parallel with a 400 μF capacitor. Calculate the line current, power and power factor when the above load is connected to a 230 V, 60 Hz, three-phase supply (2 decimal places). What power in kW would be dissipated in the load, if it is reconnected in delta (1 decimal place)?*

Reactance X_C of capacitor

$$= X_C = \frac{1}{2\pi f C} = \frac{1}{2 \times \pi \times 60 \times 400 \times 10^{-6}} = 6.63\,\Omega$$

Since the load is balanced:

The voltage across a phase, $V_{ph} = \dfrac{230}{\sqrt{3}} = 132.79\,\text{V}$

Current in the resistor $I = \dfrac{132.79}{10} = 13.3\,\text{A}$

= 13.3A, in phase with V_{ph}

Current in the capacitor $I_C = \dfrac{132.79}{6.63} = 20.03$ A

= 20.03 A, leading V_{ph} by 90°

Let I_{ph} = the resultant of 13.3 A and 20.03 A which are in quadrature

$$\therefore I_{ph} = \sqrt{13.3^2 + 20.03^2}$$

= 24.04 A. This is also the line current since the load is star connected $\therefore I = 24.04$ A

Power factor of load $=\cos\phi = \dfrac{13.3}{24.04} = 0.55$

= 0.55 (leading)

Power of load, $P = \sqrt{3}\ VI\cos\phi$

$= \sqrt{3} \times 440 \times 42.45 \times 0.55$ watts = 5.3 kW

If the load is in delta, the current per phase would be = $\sqrt{3}$ × original I_{ph}

= $\sqrt{3}$ × 24.04 A

The line current would be $\sqrt{3}$ times this new phase current.

∴ New I = $\sqrt{3} \times \sqrt{3}$ × 24.04 = 3 × 24.04 = 72.12A

The power factor of the load will remain the same

So new power, $P = \sqrt{3}\ VI \cos\phi = \sqrt{3}$ × 230 × 72.12 × 0.55 watts = 15.8 kW

OTHER BOOKS BY CHRISTOPHER LAVERS

Crosstalk- A reflection in Faith Christian poetry, October 2011 ISBN: 978-1-4709-0122-6. (Publisher: Lulu Enterprises, Inc., Rayleigh, North Carolina).
Stealth Warship Technology (ISBN 9781408175255), in the Reeds Marine Engineering and Technology series, *Adlard Cole*, Published October 2012.
Basic Electromagnetic Wave Concepts For Engineers ISBN 978-1-4709-5404-8. (Publisher: Lulu Enterprises, Inc., Rayleigh, North Carolina).
Basic Electrotechnology Reeds Vol 6: (Reed's Marine Engineering) Publisher: Adlard Coles ISBN: 0713668385 DDC: 797 Edition: Paperback; 2008-01-01 Publication date: December 2012.
Recent Developments in Remote Sensing for Human Disaster Management and Mitigation-Natural and Man-made 2013 Editor Christopher Lavers 978-1-291-22463-4 (Publisher: Lulu Enterprises, Inc., Rayleigh, North Carolina), December 2012.
EXtreme Faith Walking the Talk Various motivational Christian authors, edited by Christopher Lavers, March 2013 ISBN: 9781291351088 (Publisher: Lulu Enterprises, Inc., Rayleigh, North Carolina).
Advanced Electrotechnology Reeds Vol 7: (Reed's Marine Engineering) Publisher: Adlard Coles ISBN: 0713676841 DDC: 621.30246238 Edition: Paperback; Publication date: April 2014.
Basic Electromagnetic Wave Theory Concepts For Engineers 2014, Adlard Cole Nautical.
FICTION *TALES FROM THE FOREST SERIES:*
 # 1 ON THE WINDS OF THE DESERT SANDS November 2012, ISBN: 978-1-291-08012-4 Publisher: Lulu Enterprises, Inc., Rayleigh, North Carolina), 2012.
 # 2 ESCAPE FROM WARSAW To be printed 2013.
 # 3 TALES FROM THE WESTERN JUNGLE.

www.ingramcontent.com/pod-product-compliance
Lightning Source LLC
Chambersburg PA
CBHW072224170526
45158CB00002BA/733